Farming in Boxes

Farming in Boxes

ONE WAY TO GET STARTED GROWING THINGS

PETER and MIKE STEVENSON

CHARLES SCRIBNER'S SONS NEW YORK

CONTENTS

1

**Starting
the Farm** 6

2

Planting 24

Library of Congress Cataloging in Publication Data
Stevenson, Peter, 1941–
  Farming in boxes.
  SUMMARY: Details how crops can be grown
anywhere by raising them in wooden planter
boxes. Includes directions for building the boxes
and a simple "barn" of boards and plastic for
creating a compost pile.
  1. Container gardening—Juvenile literature. 2.
Gardening—Juvenile literature [1. Container gardening. 2. Gardening] I. Stevenson. Mike, 1965–
joint author. II. Title.
SB-418.S72   631.5   76–15383
ISBN 0-684-14673-8

3
Barn Raising 32

4
Raising Crops 50

Materials List 64

DEDICATION

For Grandma and Grandpa

We would like to thank the members of our crew for the first Box Farm:
Ricky Wilson, Chip Shore, Cece Durante, and Jeff Baldwin.

1
Starting the Farm

This is the story of how a Box Farm was made. In case you haven't heard, a Box Farm is a special kind of farm you can start up just about anywhere there's a little empty space. You can raise crops on a Box Farm if the soil's too rocky, or the soil is too sandy, or there isn't any soil at all —like on a paved lot or a corner of a playground.

One of the best things about a Box Farm is that it can all come about because of young people. With a farm like this, just about anybody can turn a flat area of cement or asphalt into a lush green patch of good-smelling, good-tasting things to eat. An old parking lot can suddenly find itself covered with tall bushes and flowers blowing in the breeze, bringing visits from bees and birds and other wildlife that never used to go near the place.

Starting up your own farm is a big undertaking, make no mistake. But it really isn't hard to accomplish if you take each job one step at a time. If you go at your own speed, the progress seems faster and before you know it, you've got a farm where you can grow big, red, tangy tomatoes; hot, tasty radishes; and even huge sunflowers.

If you like the idea of being outside and building a place to grow good things in, we'll tell you the whole story of how young people are getting together to create neighborhood farms (with hardly any help from grown-ups).

Just like in the old days when farmers got together to help each other build their barns and bring in the crops, it takes a lot of hands to get the job done—and have a lot of fun in the process. There are all sorts of projects on a Box Farm and there's plenty of challenge for everybody. You can pick and choose the projects you want to include in your own farm. And if you don't happen to need a barn/greenhouse/potting shed, then just build what you want to build. Each of the projects is self-contained, so you can make just the ones you like.

7

The idea for this Box Farm started out one day when nothing much was going on and everybody seemed to have lots of time on his hands.

Ideas were being thrown back and forth about how little there was to do and how lucky young people were who had all sorts of adventurous challenges to face every day—like those who grow up on farms, for example. There's always something interesting happening on a farm: dirt to be dug up and plowed into neat rows, plants to be planted, crops to be brought in (not to mention watermelons and other good things to be eaten).

A farm would be fun, it was agreed; but how do you start a farm where there's nothing but asphalt and buildings—grow crops in flowerpots?

Then someone remembered seeing people grow big vegetables in planter boxes in areas where the soil was poor for gardening. Why couldn't you build a bunch of planter boxes and start up a farm where there was no soil at all? Just bring in some good dirt, a little fertilizer, and water, and you'd be set.

It was all settled. First they picked out a small corner of an unused lot where the sun was good most of the day, and then got permission to use it for "an important agronomy study"—an experiment in farming—to see if they could raise crops in the middle of a city or town. Then they began planning what would be required for the farm. First they'd need boxes to mix up the right kind of soil in. Then they'd need a barn, of course. Every farm has to have a barn, and they would make theirs of inexpensive wood and plastic so it could also be a greenhouse and a potting shed where they could raise extra-delicate plants. And it could be a place to store all their tools, fertilizer, and equipment out of the weather.

Next, they'd need a compost pile. With a compost pile you can turn weeds and old leftover crops into rich fertilizing mulch for the crops. That way, old piles of plant clippings, melon rinds, leftover tomatoes, and the like wouldn't become a trash problem; they would be recycled right there to solve future mulch needs.

The first job on the project list was to get rid of any loose junk around the lot so the farm could start out with a fresh, clean area. The next challenge was to find enough wood to get the boxes built. Once the boxes are filled with soil, the first crops can be planted and will be growing while the barn, compost pile, and any other projects are carried out.

A lot of good lumber is thrown away all the time, and you can find plenty of discards that come in handy by cleaning out around the backs of alleys and the like. But to get the big boards used for the box sides and bottoms you may need to get some aid from an older helper who can lend a hand by actually buying the boards at a lumber-yard (just point out that it's an investment in garden-ripe, fresh vegetables and flowers to come).

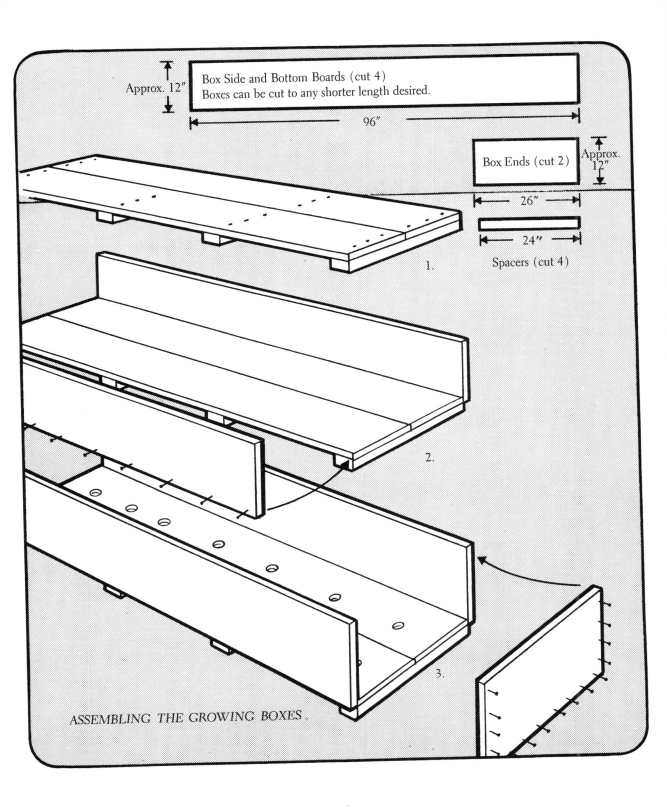

Box Side and Bottom Boards (cut 4)
Boxes can be cut to any shorter length desired.

Approx. 12"

96"

Box Ends (cut 2)

Approx. 12"

26"

24"

Spacers (cut 4)

1.

2.

3.

ASSEMBLING THE GROWING BOXES

Use the roughest, cheapest grade of redwood or cedar lumber for the boxes. These woods won't soften when filled with dirt.

The boxes are built with 2" and 3" nails (a complete list of just what you'll need for each project can be found at the back of the book). Galvanized nails hold best and don't rust. But don't put them in your mouth while nailing. New carpenters are always getting sick from holding galvanized nails in their mouths while working.

The drain holes are important because they allow the water to drain through the soil. Use the worst boards for the bottom boards. Any small (up to 1½″ diameter) knot holes in the boards can be used as extra drain holes. Larger holes can be plugged by nailing plastic or metal can tops over the holes from the inside of the box.

When sawing, try to place the board to be cut on a firm surface, held securely. The more firmly the board is held, the easier it will be to cut through it.

If it's all right for you to use an electric drill for the 1″ drain holes through the bottoms of the boxes, here are some tricks that make boring holes a lot easier.

First, make sure you have an older helper around to lend a hand when needed. Again, the more firmly you can hold the board, the easier it will be to drill. Make contact with the drill bit on the wood before pulling the trigger. And release the trigger and let the drill stop before removing the bit from the wood. If a power drill isn't available, the holes can be bored with a regular brace-and-bit, or a hand drill.

To assemble the boxes, place the 2″ x 2″ spacers under
the bottom boards, which are aligned side by side, with
their ends even. Nail down through the bottom boards
and into the spacers with 2″ nails. Drive 3″ nails just into
the side pieces, spacing the nails about 6″ apart and about
3⁄8″ to 1⁄2″ up from the bottom edge of the sides.

Hold the side up against the bottom, aligned flush at
the bottom, nail the end nails, then drive in the remaining
nails through the side and into the edge of the bottom
board. Drive 3″ nails just into the box ends, spacing the
nails about 4″ apart and 3⁄8″ in from the bottom and side
edges. Hold the end in place against the ends of the sides
and bottom and then drive the nails in through the end
piece and into the end edges of the sides and bottom.

Once you find out how easy it is to make one box, you may want to set up a production line to build all the rest of the boxes you need in a hurry. One or two builders can run the saw, while another assembles box bottoms, and still another nails and attaches sides and end pieces.

By the time the Box Farm reached this stage of the game, it began to gather curious onlookers. And somehow, before the day was out, some of these observers found themselves in the thick of things, working as hard as any of the original farmers.

From then on, the working crew was liable to change from day to day, as the farmers fitted their time on the projects into their own schedules. And soon it became a definite community effort as the members of the neighborhood added their own touches to the Box Farm.

Even though the boxes take a little extra effort to build, when the growing starts, they'll begin to pay off in the added ease of controlling soil conditions.

Ever since the first caveman decided to move some of his favorite plants a little closer to the cave so he could get at them more easily, farming has been one long battle for control over growing conditions.

If farmers could always control the make-up of the soil, have the sun shine and the rains come on schedule, and keep the bugs away, they'd never have a problem. But soils go bad, dry seasons come, and bugs attack—and farmers have their hands full trying to feed the expanding population.

Farming in boxes gives you a better chance of controlling growing conditions. The soil can be mixed in just the proportions of mulch, sand, fertilizer, and nutrients you need, like mixing up a cake. The watering can be precisely metered to the needs of the particular crops in each box, weeds can't creep in from the soil around the farm, and crawling bugs are discouraged by the high walls of the boxes set off the ground.

As soon as the hammering and sawing have drawn to a close, it will be time to figure out just where you'll want to place your crops. They'll need plenty of sun, not too much wind, and room to grow; and you'll need space to move around the boxes after the crops start shooting up and out.

It may take a bit of trial-and-error experimentation before you get things the way you like them. This crew took up half a day moving the boxes around until everybody was finally satisfied with the arrangement (or too tired to argue anymore). It's a good idea to get things just right at this point. Once the boxes are filled up, they are pretty hard to move around.

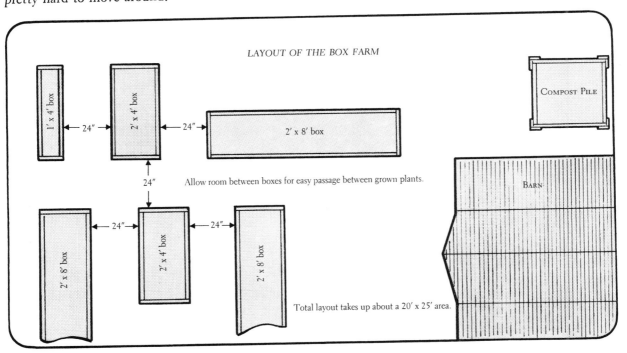

LAYOUT OF THE BOX FARM

1' x 4' box

2' x 4' box

←— 24" —→   ←— 24" —→

2' x 8' box

COMPOST PILE

↕ 24"

Allow room between boxes for easy passage between grown plants.

BARN

2' x 8' box

←— 24" —→   2' x 4' box   ←— 24" —→

2' x 8' box

Total layout takes up about a 20' x 25' area.

Once the boxes are laid out, it's time to mix up the soil and plant the first crop so that it can be growing while any other building projects are under way.

The first step is to prospect around for a source of dirt to use as a base for the soil. Then food, looseners, and water retainers can be added to create a good growing mixture.

You may be limited to just the dirt you can find nearby. But even if you find out it has drawbacks, most can be remedied by the other components of the final soil.

A quick look at the job the soil has to do can show you what kind of food and soil treatments to mix in. A good soil has to be loose enough to let the roots grow large easily and to let the water flow down to bring food to the roots (and then to flow out so that the next watering can bring new food in). Good growing soil should have plenty of food in the form of fertilizers. Some bottled plant foods have nitrogen in them to make the plant stalks and leaves grow fast, while others have phosphorus in them to make the blooms, flowers, and vegetables grow fast.

We'll go into more about soil mixing later on. For now, it's enough just to get an idea of the sort of job a good growing soil has to perform so you can scout around for a source of dirt that looks suitable.

It's a good idea to sift the dirt through a screen with holes about ½″ square to get rid of rocks and sticks as you're filling up the first loads.

A good growing soil should also have old leaves and wood chips mixed in to help hold the water so it doesn't drain through too fast.

If your soil happens to be too thick with clay, you can mix in extra looseners like leaf mulch, bark chips, or sand (not beach sand, though, because of the salt) to keep it from hardening up after watering. If it's too sandy, mix in more wood chips to hold the water. You can buy kits to test the acidity of your soil at a nursery if you really want to be precise. Then you can ask the nursery proprietor for the right kind of food needed to balance out the acidity your soil may have.

As you start ferrying the dirt in wagons or carts, bring in enough to fill your boxes up to about 2/3 to 3/4 full. If it looks to you like good growing soil, leave about 1/4 of the box empty for fertilizer/mulch mix. If it looks like the dirt is going to need a lot of help, leave the box emptier for the extra soil additives it will need.

You can mix your own fertilizer/mulch combination or buy it ready-mixed at the nursery. If you can find a pile of small leaves that has been sitting around for a while, go ahead and mix it in with the rest to add a little extra soil conditioner.

Bring the final mixture level in the boxes up to within 1″–2″ of the top.

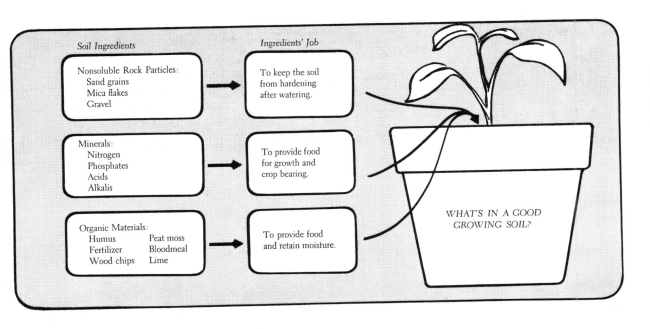

Soil Ingredients

Nonsoluble Rock Particles:
   Sand grains
   Mica flakes
   Gravel

Minerals:
   Nitrogen
   Phosphates
   Acids
   Alkalis

Organic Materials:
   Humus   Peat moss
   Fertilizer   Bloodmeal
   Wood chips Lime

Ingredients' Job

To keep the soil from hardening after watering.

To provide food for growth and crop bearing.

To provide food and retain moisture.

WHAT'S IN A GOOD GROWING SOIL?

Any soil will be a mixture of different ingredients, even though it may look like it's the same all the way through as you hold it in your hand. If we could see the small particles, we would notice some sand (small chips of rock that don't get soggy when wet), some organic matter (small bits of material that was once alive), and minerals or chemicals that may or may not mix with water to change the kind of food balance that reaches the roots.

Recently people have learned to raise plants in pure sand or even gravel by feeding the roots with a balanced diet of liquid food, using the sand or gravel only as a neutral base to hold the roots so the plants can stand up. Although certain problems have been found in using only artificial chemical food to feed the plants that are supposed to produce our own food, the experiments do show that growing things depends a lot on what you mix into the soil. And if you mix in the right proportions of natural food and fertilizer, the results can be pretty rewarding.

It's also important to get your soil well mixed through-out the box. Too much fertilizer can chemically burn the plants and hurt the growth just as much as having too little food and starving the plants. It's better to have *well-mixed* soil that's only fairly close to a good proportion than to have precisely the right proportions that aren't mixed up well.

To get the fertilizer/mulch mixed well down into the box, first dig with a shovel to turn the nutrients under the dirt, then mix up thoroughly with a hoe. It takes a bit of armwork, but will pay off at harvest time just as much as any other step in the project.

There are a few tricks to getting good crops that ought to be thought about now, even though they may not be used until later.

As we said, the nitrogen plant foods you can buy help the green parts of the plants grow, while the phosphorus foods help the blooms and vegetables grow. If you decide to try bottled plant foods, be sure to get an older helper to lend a hand in getting the measurements right. They're powerful stuff and a small mistake can do as much damage as the right proportions can help.

The plan, obviously, with most plants would be to give them a dose of the green-growing food at the start to get growing well under way. Then follow up with some food for the blooms later on, as the plants are ready to bear.

Don't forget that with some crops like lettuce and chard what you're eating is the leaf, so there's no need to add food for the blooms.

2
Planting

Once you've reached this stage of the project, you're through the hardest part. From now on, the jobs lighten up and you can start to watch your efforts pay off.

The crew of this farm decided to divide up both the seeds and the boxes so that each member had his own crop to take care of (although everyone eventually ended up taking care of everybody else's crops from time to time). To keep things straight, a map was made of the farm layout and the boxes were numbered so that a list could be kept of who had planted what where, and how each box should be fed and watered.

You have to do a little planning ahead of time to get the most out of your farm. Some plants are kept wet until they spring up above ground level, while others are watered only once before they spring up. Some crops, like tomatoes, need to be dried out to get the fruit to ripen, so these shouldn't be right next to crops that need constant watering. Tall, leafy plants shouldn't be planted where they'll shade other plants that need lots of sun.

The kind of treatment each crop needs both for planting and growing can be found on the backs of the seed packs. When the seeds for this farm were bought, the crew spent a long time "horsetrading," each trying to get his own favorite vegetables and flowers to plant in his box. Then they realized, after reading the requirements for their choices, that some had to be retraded so that the crops could be laid out in a practical way to get the best results from each box.

25

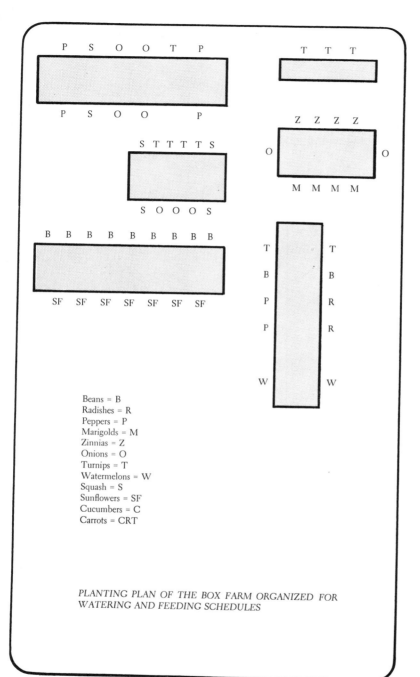

P S O O T P

T T T

P S O O P

Z Z Z Z

O   O

S T T T T S

M M M M

S O O O S

B B B B B B B B

T   T

B   B

P   R

P   R

SF SF SF SF SF SF SF

W   W

Beans = B
Radishes = R
Peppers = P
Marigolds = M
Zinnias = Z
Onions = O
Turnips = T
Watermelons = W
Squash = S
Sunflowers = SF
Cucumbers = C
Carrots = CRT

PLANTING PLAN OF THE BOX FARM ORGANIZED FOR
WATERING AND FEEDING SCHEDULES

Even though you're given a little added control by growing crops in boxes, the season and the weather in your particular area will limit what you can plant at any one time of the year. The people who run the local nurseries in your area are the most likely authorities on when to plant what to take advantage of your weather conditions.

Basically, you can choose from cool-weather crops and hot-weather crops. Some cool-weather crops that grow well are: peas, sweet peas, lettuce, broccoli, and cauliflower; ranunculas, daffodils, crocus, tulips, and snapdragons.

Some crops that grow well in warm weather are: tomatoes, squash, melons, peppers, onions, corn, carrots, cucumbers, beans, radishes, and turnips; daisies, zinnias, dahlias, petunias, and sunflowers.

Don't be shy about asking advice on how to grow crops in your particular area from the person selling the seeds. He is all for making those seeds a success and knows all the local stories and advice about the way different crops grow where you live.

The period from the planting of the seeds to the time when the plant pops up above the soil is very important to the later growth of the plant, so pay close attention to the instructions on the seed packs.

Some crops aren't usually grown from seed, however. Tomatoes and peppers are handier to grow from seedlings —little plants just started in trays. These are available, along with certain flowers that are grown from bulbs instead of seeds, at any nursery. Generally, tomatoes are planted deep so that the soil covers their stalks about halfway up to the leaves.

Because you're working with a limited area in a Box Farm, you can plant the seeds a little closer together than indicated on the seed packs. Usually, the plants will be trained upward to climb on trellises (sort of like a "high-rise" farm). Once the young plants are growing, you can thin them by pulling out a few to give the healthiest plants plenty of room to spread out.

This crew found that if they were a little careless with the watering of newly planted seeds, they could easily wash them away in the flood of water. To prevent this, you can spread a rag or paper towels over the row to be watered, then pour the water from the watering can directly onto the towels. This way the water can soak gently through without disturbing the placement of the seeds.

It's a good idea to water down the soil well with a deep soaking before planting. This way the moisture from the first watering won't be absorbed away from the seeds by the dry soil underneath.

The kind of watering style used can have an effect on how the crops grow. Even if you have a carefully arranged schedule of how much water to give each box, you have to make sure the water is reaching the roots to bring in the food and minerals.

Ten gallons of water dumped at one time into a box will tend to just soak the top few inches of soil, while the same amount of water dripped or applied in repeated waterings to a few spots on the soil surface will penetrate deep down into the soil at these points. So, you should keep in mind what the roots of each crop look like beneath the soil so that you can use the kind of watering that will reach the whole length of the roots.

In general, you can think of each plant's roots as looking pretty much like the plant itself, only upside down underneath the soil. Tall plants often have deep roots and shorter crops usually have roots nearer the surface.

The crew of this farm found that at least two passes with the watering can or hose were needed at each watering. The first soaking just wets the surface (which tends to be baked into a crust by the sun). Then the second watering has a better chance of soaking down through the looser soil underneath.

Even though the instructions on the seed packs were definite about how long it would take the seeds to push up through the soil as new plants, there was still a good deal of impatient pacing around the boxes by the crew as they waited for the first signs of life. All the pacing and arguing probably didn't really hurt anything, since some people say that talking to plants helps them grow faster.

In any case, to the delight of the farmers, the first plants started showing through the surface of the dirt long before the designated period (germination time) was up.

At first, there was some excited talk of a foul-up in the planting layout, because the baby plants —like baby anythings—all looked pretty much alike and they couldn't tell the squash from the beans.

By the following week, however (p. 30, lower left), the crops began to sprout second leaves which looked more like those of the full-grown plants—and the crops could be told apart.

By the second week (below), each crop had lost its baby leaves and was well on its way to becoming full-fledged plants.

Once the plants are big enough to recognize, serious weeding can begin. Weeds steal food and living space from the crops. And because they're usually stronger growers than the highly developed kinds of plants that produce the biggest crops, they can take over a farm and force out the plants if they're not removed. To weed safely, make sure you can recognize where your crop is, then pull everything else out.

3

Barn Raising

By the time the crops are well on their way and growing steadily, the watering and weeding settles down to an easy routine. And you can start to think about further improvements to your farm-plant facilities.

The barn built for the Box Farm was styled after the old Dutch curb-roof barns that have dotted the countryside for centuries.

Barn raising is a pastime and great social event that dates back before the beginnings of our country. Whenever a farmer needed a new barn, the word went out and neighboring families from miles around would show up in their wagons on the appointed day. With plenty of hands to help, a barn could be built amazingly fast, without taking too much time away from the farmers' daily work schedule. Barn raisings were good excuses for getting together mountains of food, enjoying the latest gossip, and showing off how much could be built in one day.

To build the barn the basic steps are to cut the 2″ x 2″s and 2″ x 4″s to the lengths shown. Then notches are cut, as indicated, in the 2″ x 4″s; the 2″ x 4″s are laid out; then the 2″ x 2″s are laid in the notches and nails are driven through the 2″ x 2″s and into the 2″ x 4″s at the notches. Each wall frame is then covered with plastic, the frames are stood up and nailed at the corners, and the roof panels are laid on top after plastic is stretched over the tops of the frames.

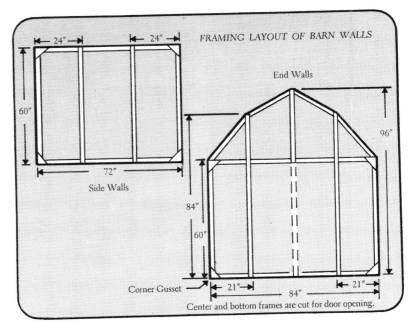

FRAMING LAYOUT OF BARN WALLS

End Walls

Side Walls

24″ 24″ 60″ 72″

84″ 60″ 96″

Corner Gusset 21″ 84″ 21″

Center and bottom frames are cut for door opening.

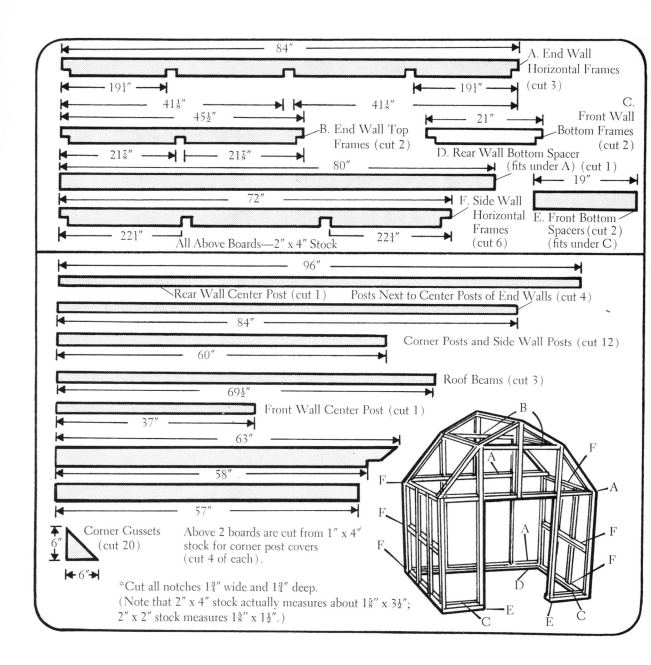

84″

A. End Wall
Horizontal Frames
(cut 3)

19⅛″   19⅛″

41⅛″   41⅛″

45½″

C.
Front Wall
Bottom Frames
(cut 2)

21″

B. End Wall Top
Frames (cut 2)

D. Rear Wall Bottom Spacer
(fits under A) (cut 1)

21⅞″   21⅞″

80″

19″

72″

F. Side Wall
Horizontal
Frames
(cut 6)

E. Front Bottom
Spacers (cut 2)
(fits under C)

22¼″   22¼″

All Above Boards—2″ x 4″ Stock

96″

Rear Wall Center Post (cut 1)   Posts Next to Center Posts of End Walls (cut 4)

84″

Corner Posts and Side Wall Posts (cut 12)

60″

Roof Beams (cut 3)

69½″

Front Wall Center Post (cut 1)

37″

63″

58″

57″

Corner Gussets
(cut 20)

6″

6″

Above 2 boards are cut from 1″ x 4″
stock for corner post covers
(cut 4 of each).

*Cut all notches 1¾″ wide and 1¾″ deep.
(Note that 2″ x 4″ stock actually measures about 1⅝″ x 3½″;
2″ x 2″ stock measures 1⅝″ x 1½″.)

DIMENSIONS FOR BARN PARTS

After cutting out the 2″ x 2″s, 2″ x 4″s, and gussets (corner braces), start out the framing by laying out one side wall. Place two 2″ x 4″s on a flat surface with their notches facing up, so that they are about 5′ apart. A third 2″ x 4″ should be placed 30″ from one of the 2″ x 4″s. Then lay four 5′ lengths of 2″ x 2″s across the notches of the 2″ x 4″s. Nail down through the 2″ x 2″s at the four outside corners, then nail the third 2″ x 4″ so that its top surface is 30″ from the bottom of whichever 2″ x 4″ you choose to be the bottom of the side wall.

To square up the sides, measure across diagonally from one corner to the far opposite corner. Then measure across the other two corners the same way. By pushing opposite corners together or pulling them away from each other, you can adjust the wall until both measurements are the same.

Then have someone stand on the wall to hold it steady, and nail the plywood corner gussets to the frame at the corners, driving at least two nails into each frame piece. This way you can square up the walls and hold them square until the plastic is nailed to the edges.

36

Lay out the rear wall after the side walls. Lay out the two 7′ lengths of 2″ x 4″s, with the notches facing up, 5′ apart. Lay 5′ lengths of 2″ x 2″s in the end notches, nail the parts together, and square up and nail corner gussets in place to hold the frame square. Next, lay the two 7′ lengths of 2″ x 2″s in the next two slots toward the center and nail in place, flush at the bottom. Place the 8′ length of 2″ x 2″ in the center notches and nail.

The front wall is framed the same way, except that the bottom of the frame is made up of two short 19½″ lengths of 2″ x 4″s notched at the ends. The center vertical 2″ x 2″ is cut to 37½″ and extends from the bottom of the full-length 2″ x 4″ stretched across above the door up through the notch in the 2″ x 4″ above.

To frame the angled roof line, lay 2" x 2"s across the top corners of the end frames, flush at the outsides of the corners. Then mark the 2" x 2"s underneath, as shown, where they sit against the frame parts. Cut these and nail through them to the frame to attach in place.

Once the four walls are framed, you can place them down on the plastic and trim around the walls about 3" out from the sides of the wall. Nail the plastic to the frame with roofing nails, fold in the plastic up along one side, and nail through the plastic to the outside edge of the frame every 6" or so.

Pull the plastic tight on the opposite side and nail. Then nail the remaining side edges, one at a time.

To attach plastic to the rear wall, nail first along one side edge, then the opposite side. Nail along the bottom next, and finally along the angled top edges. Nail the plastic to the front wall in the same way, then cut around the plastic about 3" inside the door and nail.

Lay out the end walls, as shown, to get the most from the sheet of plastic. An easy way to cut the plastic is to hold the scissors halfway open and just slide them along after the cut is started. The plastic will slice so fast this way that you have to be careful not to wander away from the line you want to follow.

39

The slat roof covering the barn is made up of four panels: two upper panels and two lowers. The frames of the lower panels are cut from 1″ x 2″ stock. To assemble the lower panels' frames, lay one of the long 1″ x 2″s on edge and nail through one of the shorter 1″ x 2″s and into the end of the longer board to attach them at the corner (with both boards on edge). Attach the other short 1″ x 2″ to the other end of the long 1″ x 2″. Then nail the remaining long 1″ x 2″ between the other ends of the short 1″ x 2″s, but place this longer 1″ x 2″ on its side, instead of on edge, and align the top side flush with the top edges of the short-end 1″ x 2″s before nailing through the end boards and into the end grain of the long 1″ x 2″. This 1″ x 2″ will form the top edge of the lower panels.

To form the upper panels, lay the longer 1″ x 2″ up against the side of the shorter 1″ x 2″, with an equal amount of the longer 1″ x 2″ sticking out on both ends. Mark the position of the ends of the shorter board on the longer one, then nail 25″ slats from the shorter board across to the longer, so that the outsides of the slats are lying along the ends of the shorter board and the marks on the longer. The 1″ x 2″ end frames will be cut to fit and attached to the ends of the top panel frames after the slats have been nailed to the frames and the frames laid in place on top of the barn.

Slats are nailed across the framework of the roof panels. Because there are so many of them, it's a good idea to rig up a simple jig so you can cut several at a time without having to remeasure each time you cut.

To set up a jig, simply drive two nails down into the bench used as a cutting table to hold the slats on the side. Then place a third nail so that when a slat is held against the first nails, the third nail will sit against one end of the slat, 36″ from the end of the bench (for the lower panel slats), and 25″ from the end of the bench (for the upper panel slats).

You can place three or four long slats in the jig and cut them off at the end of the bench to set up a fast production line of slats cut to fit. Use the slats themselves to space out the slats to be nailed to the frames. Nail the first slat to the end of the panel frame. Then place a second slat down against the first slat, and then a third up against the second. Nail the third slat to the frame. Then place two more slats up against the last slat nailed, skip the middle slat and nail the last slat placed, and so on across until the whole frame is nailed with slats.

Place a short scrap of board under the upper side of the lower frame to brace it when nailing on the slats. When the last slat is nailed, the barn kit is complete and you're ready to start the actual assembly.

The first step of the barn raising is to lift up the rear wall and prop it against something solid. Next, a side wall can be raised and leaned against something near the rear wall. Bring the end of the side wall up to the edge of the rear wall, then lift the rear wall a couple of inches so that you can bring the edges together, with the bottom 2″ x 4″ of the rear wall resting on the bottom 2″ x 4″ of the side wall. The upper 2″ x 4″ of the rear wall should also be resting on the upper 2″ x 4″ of the side wall. Nail down through the 2″ x 4″s where they overlap at the corners to secure the corners.

If the barn is to be removed for storage at a later date, it can be knocked down easily by removing these nails (so don't drive the nails all the way in for easier removal later, if desired). At each corner, the 2″ x 4″s of the front and rear walls rest on the 2″ x 4″ of the side walls. Cut 2″ x 4″s to fill in the gap between the bottoms of the front and rear walls and the ground, and drive nails down through the bottom 2″ x 4″s and into the 2″ x 4″ spacers you just cut to fix them in place.

After the second side wall has been joined to the rear wall, raise and attach the front wall to the front edges of the side walls, resting the 2" x 4"s of the front wall on the 2" x 4"s of the side walls and nailing through.

With all four walls standing, measure across diagonally inside the barn from corner to opposite corner and adjust the positioning of the walls until the barn is squared, just as the wall frames were squared up.

Cut a 2" x 2" to fit between the front and rear wall at the peak of the roof, and attach by nailing in through the wall frames at the top and into the ends of the 2" x 2".

Cut and attach 2" x 2"s between the front and rear walls at the break in the roof angle line on both sides of the peak. These 2" x 2"s will actually extend between the ends of the upper horizontal 2" x 4"s. The ends of the 2" x 2"s can be attached to the ends of the 2" x 4"s by placing a plywood corner gusset over both boards and nailing down through the plywood in both frame pieces, as shown here.

44

To provide watertight protection for the farm equipment, pull the remaining plastic over the top of the barn. Nail this along the top edge of one side wall, using roofing nails, then pull it tight from the other side and nail along the top of the opposite wall. Nail across the top of the rear wall, then pull it tight from the front and nail across the top of the front. Trim away excess plastic with the scissors, and you're ready to attach the roof panels.

To attach the lower roof panels, hold them in place, as shown, with the flat, mounted 1″ x 2″ to the top. Nail down through this and into the 2″ x 2″ roof beam at the break in the roof line. The top panels can simply lie flat on the barn top and rest at their bottom edges on the top edges of the lower panels. If you need a higher temperature inside the greenhouse/barn for raising potted plants in cold weather, simply remove the upper roof panels to let in more light and add a simple plastic flap across the door to keep out cold breezes.

Mark and cut 1″ x 2″s on the angles indicated to mount the end trim to the upper roof panels. Then nail leftover slats in place, flush at the tops of the frames; trim off flush at the front of the frames.

To make built-in workbenches for the barn, simply cut wide scrap boards to fit between the front and rear walls; nail to the top of the middle horizontal 2" x 4" of each wall and support the front edges with 2" x 2" or 2" x 4" legs, nailed under the front of the bench.

If two boards are to be joined side by side to make wider benches, nail boards across beneath the boards (as with the spacers under the box bottoms). Then mount the assembled bench top in place and nail front legs under the bench front, up against the cross-boards.

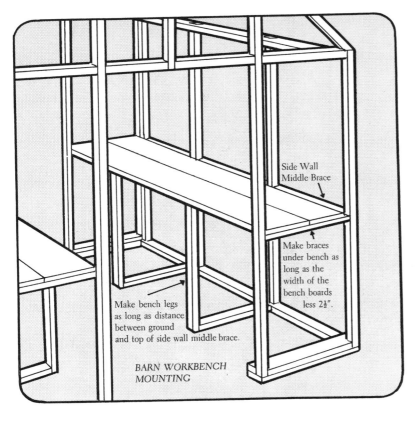

Side Wall
Middle Brace

Make braces under bench as long as the width of the bench boards less 2½".

Make bench legs as long as distance between ground and top of side wall middle brace.

BARN WORKBENCH
MOUNTING

To cover the gap at the corner joints of the walls on the outside, cut 1" x 4"s, as shown, and join together by nailing through the angled 1" x 4" and into the shorter 1" x 4", with both boards aligned at the bottom.

Note that the right-hand and left-hand corner post covers will be assembled oppositely, so make up one set of corner post covers for each corner of the barn (instead of mass-producing these). Then simply place up against the corners of the barn and nail through into the corner posts.

Since the days of the first barn raisings, it's always been good luck to top off a new barn with a "hex sign." Hex signs are said to keep away bad luck, storms, floods, and evil spirits in general. Of course, we all know this is just a superstition, but then again, it never hurts to be on the safe side. So a special hex sign was made up for the Box Farm by cutting a circle out of a scrap of plywood and covering this with an improvised design cut from colored Contact paper. Each barn has its own sign, so you can make up any shape and design you want and nail it to the front of the barn just above the door.

At this stage of any good barn raising, it's time to start putting down the tools and start bringing out the refreshments, which take on an extra tastiness after you've just completed a job well-done and deserve a good break.

A successful barn raising is always reason for celebration, and it's even more rewarding when you realize that the crops have been busily growing all during the building. With the plants growing higher every day, you should start thinking about thinning out the smaller plants, transplanting, and building climbers so the crops can be trained up vertically. "Highrise" trellises let you grow the largest amount of crops in the smallest area of ground.

4
Raising Crops

It's a pretty lucky farmer who has all his seeds come up. Usually there will be a few blank spaces and a few overcrowded areas. Transplanting the young plants before they get too large will help even things out in the boxes. Prepare the hole in the area to be filled and give it a preliminary watering; then carefully dig out the plant to be moved, leaving plenty of soil around the roots so they aren't disturbed. Rewater the plant after the move has been made, and it will be well on its way in the new location.

For climbing plants that were seeded in a clump or hill, you can drive stakes in around the plants to make a "bean tower," as the crew called it. Nail scrap spacers between the opposing stakes, or tie them together at the top, teepee fashion. Then twine can be tied to the bottom of one stake and wound around and up the tower to give the feelers of the plant something to grab onto.

Use the twine to tie the top of the tower to nails driven into the box, making "guy wires." These will brace the tower and keep it from blowing over when weighted down with a heavily laden bush.

As you begin the second stage of the farming, you may start to see that your prized plants are attracting some unwanted fans: the bugs.

It can be downright disheartening to stand by and watch as intruders fly in from miles around to sample the results of all your efforts—even before your crops start to bloom.

As we noted earlier, snails and other crawling bugs are kept down to a minimum pretty well by the walls of the boxes, but it's harder to keep out airborne attackers. Daily attention to discovering and destroying all possible bugs helps to a certain extent. And smaller plants can be protected for a while by wire cages. But this gets harder as the plants grow larger.

You may want to resort to chemical insecticides that you can buy. In practice, these don't seem to be as foolproof as the advertisements might lead you to believe. Even after spraying, you'll still have a few bugs. So, keep this in mind if you can't quite make up your mind about the insecticide issue.

Be certain to have an older helper lend a hand if you do decide to spray, and read the warnings about not eating any of the crops too soon after spraying.

Long rows of beans can be trained upwards with a trellis built along the side of the box. To attach a trellis to the boxes, start out by nailing 2″ x 2″ end poles firmly to the ends of the boxes. Next, nail another pole across the tops of the uprights, then drive nails partway into the sides of the box next to the bean row, spacing the nails out about 4″ apart and about 1″ from the top of the box.

Tie the end of the twine onto the end nail, lead the twine up over the top pole and back down to the next nail, and so on across the whole trellis.

It's amazing how fast the climbing plants will catch on that there's something for them to climb and will start reaching out for it immediately. With no outside help, the feeler in the photo below had already wrapped itself around the twine and was heading upward like something out of JACK AND THE BEANSTALK in less than one day after the trellis was installed.

53

There's more to harvesting than just picking beans off the vine. In fact, knowing how to harvest can be one of the most important steps from seed pack to dinner table, and is one of the secrets of why vegetables that you grow yourself taste so much better than those from the store.

The big farms that raise crops for markets have to take a lot of things into consideration besides flavor. One is how the crops will transport to market; another is how they'll stay fresh until sold; and a third is selling the crops when the price is right. As a result, harvesting can be rushed with some crops picked while still green, in the hope that they will ripen on the way to market.

Most commercial crops *do* turn the right color, but few have the same full taste of those grown on a Box Farm that can ripen on the bush and then be picked just at the time when they taste the best. Being able to guess when a crop will taste the best by looking at it is an art that takes a little experience to perfect.

One method of making your crop appraisal a little more foolproof is to set aside a small portion of the harvest for testing. When the crops look about ready, you can pull out or pick samples from your test plants to see how the taste is progressing. Generally, crops start out sour and strong, then sweeten up as they ripen and the sugar content builds up. If the taste is too strong or sour, let the crop ripen a while longer and test again. Crops that are allowed to overripen may tend to lose taste and become woody, so if the test harvest is a little low on flavor, you'd better start picking right away because the taste isn't likely to build back up again.

In most cases, you can count on radishes coming up first (in fact, they'll come up so fast they may be ripe before you're ready for them). Tomatoes and melons will be the slowest to ripen, with the greens and vine plants falling in between.

Tomatoes take a little special care to bring to full ripeness. Farmers call the treatment "putting the plants under stress." This simply means that once the tomatoes grow to about the right size for picking, the watering of the plant is cut down to little or nothing. When this happens, the plant throws all its energies into the fruit itself, instead of the green bushy part—and you end up with big, rich, red, tasty tomatoes. This may seem kind of hard on the plant, but otherwise you'll end up with beautiful green bushes and no red tomatoes.

So, when first laying out the planter boxes, try to position the tomatoes where they can be isolated and put under stress without hurting the other crops.

Because some crops ripen faster than others, you're obviously going to be all through with some plants while those next to them are just getting under way. This means either replanting to use the newly emptied space to bring in a second crop, or filling the space with transplants from an overcrowded area, according to your own personal plan.

As crops are harvested and taken out, you're going to find yourself with a lot of loose greenery on your hands. It's amazing how much can grow out of just these boxes, by mixing together a little soil, water, and sunlight.

The problem of what to do with all the residue can become the solution to where to get mulch for the next crop: by putting a compost pile to work.

A compost pile is little more than a "layer cake" made by starting with a base of soil; adding about a 4″–6″ layer of old, dead plants, overripe crops, melon rinds, and the like (the smaller the old greenery is chopped up, the better); and then adding a little fertilizer or lime to get things started off right. Repeat the layers and start the whole process over again.

If you sprinkle each layer with water and keep the pile moist as you gradually add to it over the weeks, a chemical reaction will start up which changes the old cast-off plants and vegetable table scraps into rich nutrients and mulch that can be dug out later and mixed in to enrich the soil in the boxes for later crops.

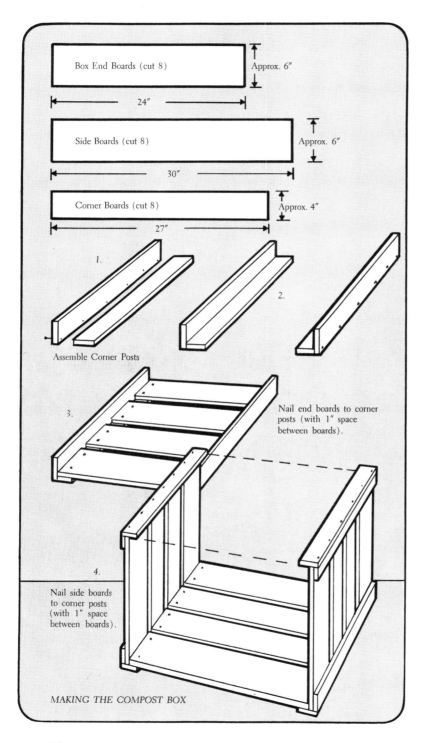

Box End Boards (cut 8)

Approx. 6"

24"

Side Boards (cut 8)

Approx. 6"

30"

Corner Boards (cut 8)

Approx. 4"

27"

1.

2.

Assemble Corner Posts

3.

Nail end boards to corner posts (with 1" space between boards).

4.

Nail side boards to corner posts (with 1" space between boards).

MAKING THE COMPOST BOX

To build a box for holding the compost (which also lets it "breathe," to keep the chemical reaction going), start out by cutting the 1" x 3" boards for the corner posts, as shown. Next, nail these together in pairs, then cut the 1" x 6" side and end boards and nail the end boards to the inside of the angles made by the corner post assemblies, as indicated.

Stand the completed ends on their side edges and then nail the side boards of one side to the inside of the corner posts on the ground. Turn the whole assembly over and nail the remaining side boards to the inside of the opposite corner posts. Then place the four sides upright in position.

Old coffee grounds, orange and grapefruit skins, and just about any kind of vegetable table scraps from home can add potency to the mulch in your compost pile. Be sure to keep the mixture moist so that the chemical reaction can take place. Heat is usually given off by the reaction and sometimes compost piles get so hot you can see steam rising from them. If you have access to a shredder, or even a power lawn mower with a collecting bag, to chop the greenery up before you put it into the pile, the reaction will happen a lot faster and you'll have your new mulch supply sooner.

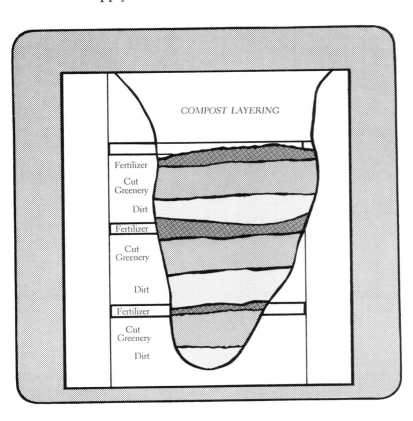

COMPOST LAYERING

Fertilizer

Cut Greenery

Dirt

Fertilizer

Cut Greenery

Dirt

Fertilizer

Cut Greenery

Dirt

By the time the farm reaches full production it may
seem a little hard to remember back when the scene of
action was just a flat, barren piece of pavement. Many
events and dramas have been played out on the area: the
tilling, planting the first crops, watching them grow up, the
barn raising, and the harvesting.

And as the flurry of activity begins to taper off with the
season gradually drawing to a close, the time eventually
comes when the old plants must be pulled out and the soil
prepared for future crops in the next growing season.

As much soil as possible is knocked out of the roots of the old plants to keep the dirt level from sinking too much as the old plants are pulled and sent to the compost pile. The boxes can be filled back up to the previous level with another dose of fertilizer/mulch mix or compost to recondition the soil for the next crops.

The farm may seem strangely quiet after the last plants have been pulled. Only a couple of puzzled bees circling around and wondering where their favorite plants have gone break the new stillness. Quite a change from all the hammering and digging and debating that went on a while back.

But with any luck, it'll all happen again pretty soon—the dirt will fly, the bushes will be blowing in the breeze, and the bees will have their work cut out for them again—not to mention the farmers.

The sun and soil have given us some good things to eat and colorful things to brighten up the place with. Time now to tidy up the place a bit, put the tools away for a while, and give the soil a little rest—time to let it store up energy for the crops to come.

Materials List

AMOUNT   LENGTH   MATERIAL

PLANTER BOXES
For each 2' x 8' planter box:

4	8'	1" x 12" rough redwood or cedar (outdoor planter box grade)
1	5'	1" x 12" rough redwood or cedar (outdoor planter box grade)
1	8'	2" x 4" or 2" x 3" or 2" x 2" (any grade)

Add one 10' 2" x 2" and one 8' 2" x 2" for each growing trellis. One ball of ordinary twine will be needed for stringing the trellises.

COMPOST PILE BOX

2	8'	1" x 6" rough redwood or cedar (outdoor planter box grade)
2	10'	1" x 6" rough redwood or cedar (outdoor planter box grade)
2	10'	1" x 4" rough redwood or cedar (outdoor planter box grade)

THE BARN

2	8'	2" x 4" construction grade, any wood
4	7'	2" x 4" construction grade, any wood
6	6'	2" x 4" construction grade, any wood
5	8'	2" x 2" construction grade, any wood
4	7'	2" x 2" construction grade, any wood
3	6'	2" x 2" construction grade, any wood
12	5'	2" x 2" construction grade, any wood
4	6'	1" x 4" construction grade, any wood
4	5'	1" x 4" construction grade, any wood
1		1" x 3" plywood scrap, any grade and thickness
1	sheet	12' x 30', 6 mil polyethylene plastic sheet (sometimes called "visqueen")
30	6'	¼" x 1⅜" redwood slats
20	7'	¼" x 1⅜" redwood slats
6	7'	1" x 2" redwood
2	8'	1" x 2" redwood
4	6'	1" x 2" redwood

NAILS (to complete all projects)

1 pound	3" galvanized box nails
1 pound	1½" galvanized box nails
5 pounds	2" galvanized box nails
1 pound	roofing nails

631.5
STE
   Stevenson, Peter
   Farming in boxes

DATE DUE			

DEMCO

DEMCO